SRA
Connecting Math Concepts

Level A Student Assessment Book

COMPREHENSIVE EDITION

A DIRECT INSTRUCTION PROGRAM

McGraw Hill Education

Bothell, WA • Chicago, IL • Columbus, OH • New York, NY

MHEonline.com

McGraw Hill **Education**

Send all inquiries to:
McGraw-Hill Education
4400 Easton Commons
Columbus, OH 43219

ISBN: 978-0-02-103595-3
MHID: 0-02-103595-4

Printed in the United States of America.

18 LON 23

The *McGraw-Hill* Companies

Mastery Test 2

Name _____

Group-Administered Scoring Section

Part 1 – 7 points each
Making Lines

Answers

Total: ⟋ 21

Individually Administered Scoring Section

Part 2 – 3 points each
What's the next number?

| Answers | 12 | 8 | 11 | 9 |

Total: ⟋ 12

Part 3 – 3 points each
Next number

| Answers | 5 | 3 | 8 | 6 |

Total: ⟋ 12

Part 4 – 4 points each
Identify symbols

| Answers | 5 | 7 | 3 | + |

Total: ⟋ 16

Part 5 – 9 points, 1st try, 4 points 2nd try
Counting

| 1 2 3 4 5 6 7 8 9 10 11 12 13 | 1st |
| 1 2 3 4 5 6 7 8 9 10 11 12 13 | 2nd |

Total: ⟋ 9

Part 6 – 4 points each

| 5 lines | 4 lines | 7 lines |

Total: ⟋ 12

Part 7 – 9 points each

How many
in the 1st group?

Answer 3
2 points

Answer 4
2 points

Get it going
& Counting

| threee 4 5 6 7 8 | 1st try 5 points |
| threee 4 5 6 7 8 | 2nd try 3 points |

| fouuur 5 6 7 | 1st try 5 points |
| fouuur 5 6 7 | 2nd try 3 points |

How many
in both groups?

8
2 points

7
2 points

Total: ⟋ 18

Test 2 Total: ⟋ 100

Remedies

Name _____

(A) Test 2 Part 1 (Lesson 18, Ex. 9)

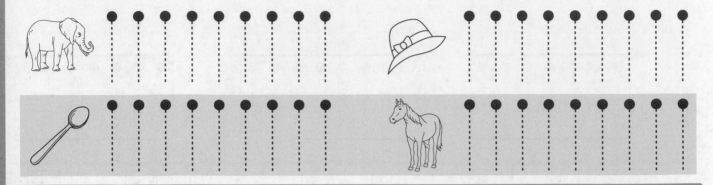

(B) Test 2 Part 1 (Lesson 19, Ex. 10)

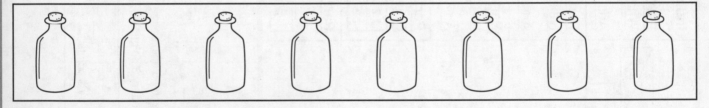

(C) Test 2 Part 6 (Lesson 16, Ex. 8)

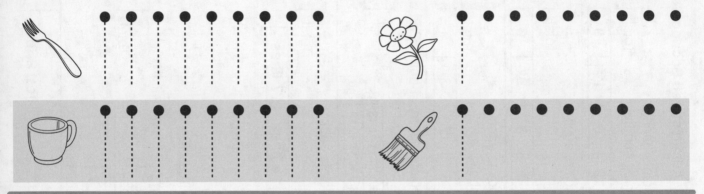

(D) Test 2 Part 6 (Lesson 17, Ex. 11)

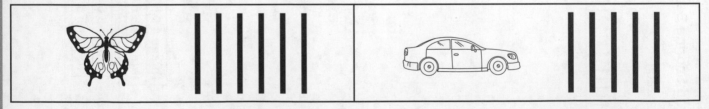

(E) Test 2 Part 6 (Lesson 18, Ex. 10)

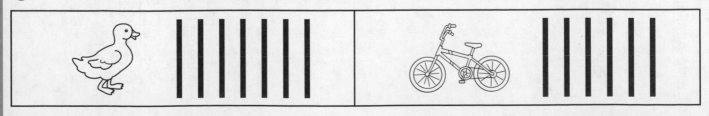

Remedies

Copyright © The McGraw-Hill Companies, Inc.

Mastery Test 3

Name _____

4 5 6 7

☐   ☐   ☐
||||| |||| |||||

4 • • • • • • •   ☐ • • • • • • •   3 • • • • • • •

🍴 |||||| = 3 │ 5 = |||||| │ ||| = 4

2 = ||| │ 6 = |||||| │ |||| = 5

🚚 6 + 2 • • • • • •   7 + 3 • • • • • •

Remedies

Name _____

A Test 3 Part 1 (Lesson 25, Ex. 6)

B Test 3 Part 1 (Lesson 28, Ex. 8)

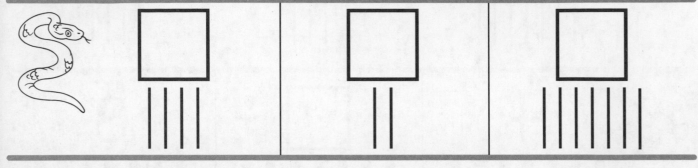

C Test 3 Part 2 (Lesson 22, Ex. 10)

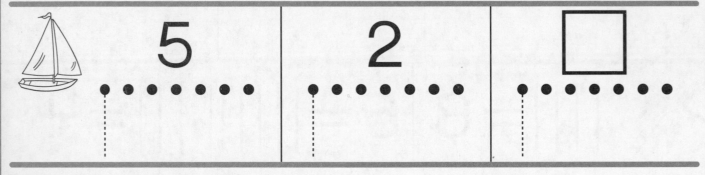

D Test 3 Part 2 (Lesson 25, Ex. 9)

E Test 3 Part 3 (Lesson 30, Ex. 11)

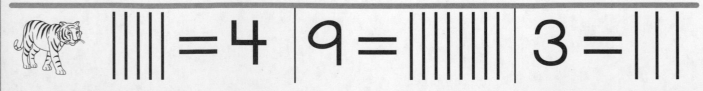

Connecting Math Concepts

Remedies CONTINUED

(F) Test 3 Part 4 (Lesson 27, Ex. 8)

 $5 + 3$ •••••• | $7 + 2$ ••••••

(G) Test 3 Part 7 (Lesson 28, Ex. 9)

 $9 + .2$ | $4 + .3$

(H) Test 3 Part 7 (Lesson 29, Ex. 8)

 $9 + .3$ | $10 + .4$

(I) Test 3 Part 8 (Lesson 30, Ex. 10)

 $3 + .4$ | $6 + .2$

Remedies

Mastery Test 4

Name _____

 ||||| • ||||||| |||||

1 + 4 = •   7 + 2 = •

 |||||||| • = • ||||| • = •

  _____

🐄 • = ||||| • = |||| | = •

|| = • |||||||| • = •   • = ||||||

🍴 6 = •   • = 3 | 5 = •

Mastery Test 4

Independent Seatwork

3   8   □   □

$6 = |||||||$ $5 = ||||||$ $|||||| = 8$

$|||||||||| = 9$ $||| = 3$ $14 = ||||||||||||||$

Connecting Math Concepts

Remedies

(A) Test 4 Part 1 (Lesson 38, Ex. 7)

||||| • ||||| •

(B) Test 4 Part 1 (Lesson 39, Ex. 7)

||||| • ||||| • |||||

(C) Test 4 Part 2 (Lesson 34, Ex. 10)

$$5 + 2 =$$ $$6 + 3 =$$

(D) Test 4 Part 2 (Lesson 38, Ex. 10)

$$3 + 2$$

Steps h–i

(E) Test 4 Part 3 (Lesson 39, Ex. 9)

||||||| • = ||||| • =

(F) Test 4 Part 3 (Lesson 40, Ex. 6)

||||||||| • = • ||||| = •

Remedies

Remedies CONTINUED

Remedies

(G) Test 4 Part 4 (Lesson 35, Ex. 11)

Steps a–c

_____

(H) Test 4 Part 4 (Lesson 38, Ex. 10)

•

_____

Steps a–d

(I) Test 4 Part 5 (Lesson 37, Ex. 10)

    |||||||| = • | • = |||| | •' = |||||

(J) Test 4 Part 5 (Lesson 38, Ex. 11)

 •   = || | |||||||| = •' • | = ||||||

Steps a–d

(K) Test 4 Part 6 (Lesson 31, Ex. 8)

 $6 = •$ | $• = 4$ $5 = •$

(L) Test 4 Part 6 (Lesson 33, Ex. 9)

 • $= 8$ $2 = •$ | $• = 3$

Mastery Test 5

Name _____

_____ | _____

 $- 5 =$ 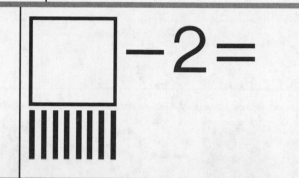 $- 2 =$

$$13 + 5 =$$

$6 + 3 =$ | $6 + 2 =$

$3 + 1 =$ | $3 + 2 =$

Mastery Test 5

Independent Seatwork

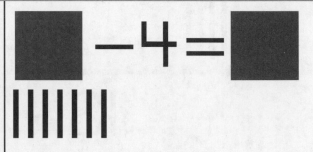

$\boxed{} - 3 = \boxed{}$ |||||||||||

$\boxed{} - 4 = \boxed{}$ |||||||

|||| =

卌 =

|||| 卌 =

= |||||||||| 卌

|||||||| 卄 =

||| 卌 =

= 3   5 =   = 2

| =   = 7   = 4

Remedies

Name _____

(A) Test 5 Part 1 (Lesson 49, Ex. 10)

Step a	Step e
_____	_____

(B) Test 5 Part 2 (Lesson 43, Ex. 7)

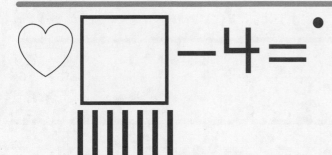 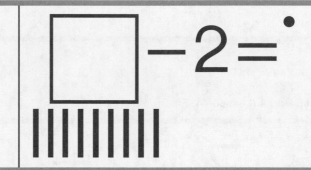

(C) Test 5 Part 2 (Lesson 47, Ex. 7)

Remedies CONTINUED

Remedies

D Test 5 Part 3 (Lesson 48, Ex. 7)

 $17 + 2 =$

$1 + 4 =$

$13 + 5 =$

E Test 5 Part 4 (Lesson 44, Ex. 8)

 $7 + 2 =$

$6 + 1 =$

Mastery Test 6

Name _____

_____

10+1=	13+1=	8+1=
6+1=	9+1=	17+1=

_____

6 – 1 =	5 – 3 =
6 – 5 =	8 + 2 =
5 + 3 =	7 – 2 =

Mastery Test 6

Independent Seatwork

$$\square - 7 =$$

| | | | | | | | | |

$$\square - 3 =$$

| | | | | | | |

$$12 + \square =$$

| | | |

$$8 + \square =$$

| | |

| | | | | =

= | | | | + | | =

| | | | | | | | | + =

| | | | + =

= | | | | | | +

Connecting Math Concepts

Remedies

(A) Test 6 Part 1 (Lesson 56, Ex. 9)

_____

(B) Test 6 Part 2 (Lesson 56, Ex. 7)

$8 + 1 =$	$10 + 1 =$
$17 + 1 =$	$6 + 1 =$

(C) Test 6 Part 3 (Lesson 55, Ex. 9)

_____

(D) Test 6 Part 4 (Lesson 51, Ex. 8)

 $7 - 3 =$ $6 - 5 =$

Remedies

Remedies CONTINUED

(E) Test 6 Part 4 (Lesson 53, Ex. 8)

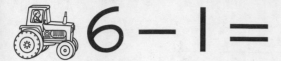 6 – 1 = | 5 – 3 =

(F) Test 6 Part 5 (Lesson 59, Ex. 8)

 12 + 3 = | 6 – 5 =

Cumulative Test 1 page 1 Name _____

   2 3 4 5 6 7 8 9 + − =

⬜ ⬜ ⬜ ⬜ ⬜ ⬜ ⬜

3 = • | • =6 | 4 = •

• = |||||||| |||||= • • = ||

3+2=   6+3=

||||||• = • |||||• = • ||| = •

Connecting Math Concepts   Cumulative Test 1   **19**

Copyright © The McGraw-Hill Companies, Inc.

$\boxed{} - 4 =$

||||•

$\boxed{} - 5 =$

|||||||•

$8 + 1 =$

$10 + 1 =$

$6 + 1 =$

$9 + 1 =$

$13 + 1 =$

$7 - 4 =$

$5 - 3 =$

Connecting Math Concepts

_____ | _____

$$7 + 2 =$$

$$7 - 2 =$$

$$6 - 5 =$$

$$12 + 3 =$$

Mastery Test 7

Name _____

$$41 + \boxed{} = 45 \qquad 8 + \boxed{} = 13$$

40	20	50

_____ | _____

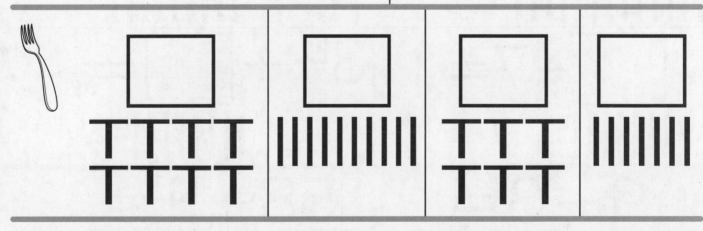

   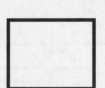 

Mastery Test 7

Independent Seatwork

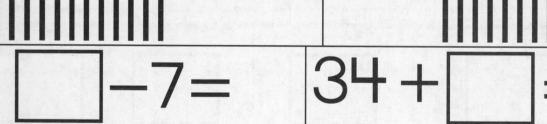

$$16 + \boxed{} =$$

$$\boxed{} - 5 =$$

$$\boxed{} - 6 =$$

$$42 + \boxed{} =$$

$$\boxed{} - 7 =$$

$$34 + \boxed{} =$$

$$9 - 3 =$$

$$58 + 4 =$$

$$7 + 5 =$$

$$8 - 2 =$$

Connecting Math Concepts

Rememdies

Name _____

A Test 7 Part 1 (Lesson 68, Ex. 8)

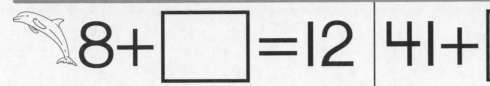 $8 + \boxed{} = 12$ | $41 + \boxed{} = 46$

B Test 7 Part 1 (Lesson 69, Ex. 7)

 $7 + \boxed{} = 12$ | $36 + \boxed{} = 40$

C Test 7 Part 2 (Lesson 67, Ex. 10)

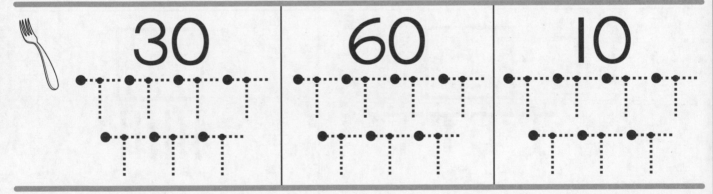

| 30 | 60 | 10 |

D Test 7 Part 2 (Lesson 69, Ex. 8)

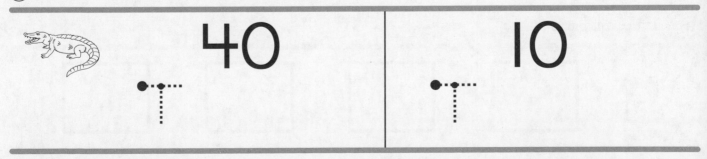

| 40 | 10 |

Connecting Math Concepts

Copyright © The McGraw-Hill Companies, Inc.

Remedies

Remedies CONTINUED

E Test 7 Part 3 (Lesson 69, Ex. 9)

_____ | _____

F Test 7 Part 4 (Lesson 68, Ex. 9)

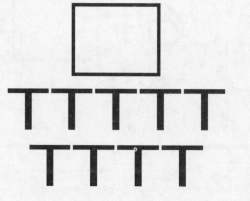 

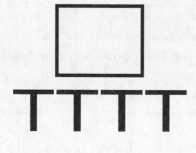 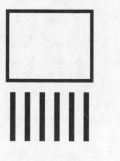

G Test 7 Part 5 (Lesson 64, Ex. 10)

Mastery Test 8

Name _____

$6 + \boxed{} = 10$

$3 + 5 = \boxed{}$

$5 + 4 = \boxed{}$

$2 + \boxed{} = 7$

 =

 =

$\boxed{} + \boxed{} = 71$

$\boxed{} + \boxed{} = 24$

$\boxed{} + \boxed{} = 35$

· 14   · 45   · 26

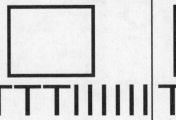
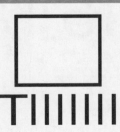
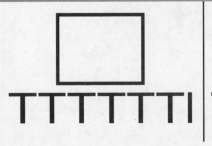
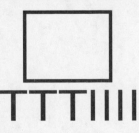

Mastery Test 8

Independent Seatwork

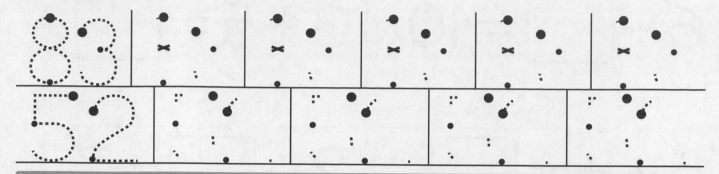

$$\boxed{} - 6 =$$
||||||||

$$\boxed{} - 2 =$$
|||||||

$$28 + \boxed{} =$$
||||

$$35 + \boxed{} =$$
|||||

$18 + 1 =$	$9 + 0 =$	$32 - 0 =$
$46 + 1 =$	$19 + 1 =$	$54 + 0 =$

$$7 - 4 =$$

$$16 + 5 =$$

Remedies

Name _____

A) Test 8 Part 1 (Lesson 78, Ex. 6)

$$3 + 5 = \boxed{}$$

$$2 + \boxed{} = 6$$

B) Test 8 Part 1 (Lesson 80, Ex. 7)

$$4 + \boxed{} = 7$$

$$13 + 4 = \boxed{}$$

$$19 + \boxed{} = 24$$

C) Test 8 Part 2 (Lesson 78, Ex. 7)

Remedies CONTINUED

(D) Test 8 Part 2 (Lesson 80, Ex. 6)

  =    =

 =

(E) Test 8 Part 3 (Lesson 76, Ex. 7)

 $\square + \square = 42$ $\square + \square = 37$

(F) Test 8 Part 3 (Lesson 79, Ex. 9)

$\square + \square = 71$ $\square + \square = 24$

(G) Test 8 Part 4 (Lesson 79, Ex. 8)

.26   .14

Remedies CONTINUED

(H) Test 8 Part 4 (Lesson 80, Ex. 8)

. 24 | . 45

(I) Test 8 Part 5 (Lesson 78, Ex. 10)

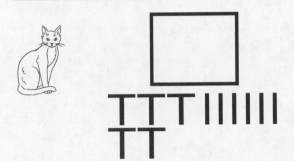

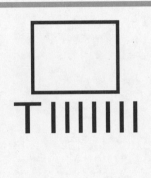

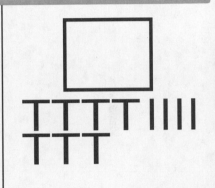

(J) Test 8 Part 5 (Lesson 79, Ex. 11)

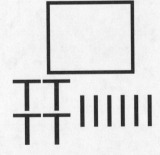

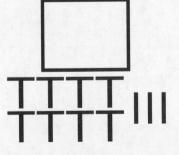

Connecting Math Concepts

Mastery Test 9

Name _____

$16 + \boxed{} = 20$

$8 - 2 = \boxed{}$

$6 + 3 = \boxed{}$

$7 + \boxed{} = 11$

21 12	72 28	15 17	64 46

$$\begin{array}{r} 14 \\ + 2 \\ \hline 16 \end{array} \qquad \begin{array}{r} 35 \\ + 1 \\ \hline 36 \end{array} \qquad \begin{array}{r} 60 \\ + 8 \\ \hline 68 \end{array} \qquad \begin{array}{r} 18 \\ - 1 \\ \hline 17 \end{array} \qquad \begin{array}{r} 27 \\ + 10 \\ \hline 37 \end{array} \qquad \begin{array}{r} 50 \\ + 2 \\ \hline 52 \end{array}$$

Mastery Test 9

$32+5=$
IIIII

$28+30=$
TTT

$32+40=$
TTTT

$28+4=$
IIII

$20+7=$ ☐

☐ $+$ ☐ $=58$

☐ $+$ ☐ $=91$

Independent Seatwork

☐ TTIIIIIIIII

☐ TIIIII

24

☐ TTTTTTIII

Remedies

Name _____

(A) Test 9 Part 1 (Lesson 83, Ex. 6)

16 + ☐ = 20   6 − 4 = ☐

37 + 4 = ☐

(B) Test 9 Part 1 (Lesson 84, Ex. 7)

38 + 4 = ☐   23 + ☐ = 25

8 − 2 = ☐

(C) Test 9 Part 2 (Lesson 83, Ex. 5)

 6 9 | 21 20 | 79 77

(D) Test 9 Part 2 (Lesson 88, Ex. 7)

106 99 | 153 150 | 64 46

Remedies CONTINUED

(E) Test 9 Part 2 (Lesson 89, Ex. 9)

| 17 15 | 12 21 | 180 108 |

(F) Test 9 Part 3 (Lesson 88, Ex. 8)

40	10	72	19	30
+6	+8	+1	+1	+5
46	18	73	20	35

(G) Test 9 Part 3 (Lesson 89, Ex. 8)

70	42	29	50	28
+6	−0	+1	+3	+0
76	42	30	53	28

(H) Test 9 Part 4 (Lesson 83, Ex. 7)

Remedies CONTINUED

Ⓘ Test 9 Part 4 (Lesson 84, Ex. 10)

Ⓙ Test 9 Part 5 (Lesson 81, Ex. 7)

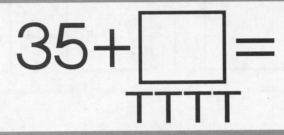

$$35 + \boxed{} =$$
TTTT

Ⓚ Test 9 Part 5 (Lesson 86, Ex. 4)

$$67 + 20 = \boxed{} \qquad 67 + 4 = \boxed{}$$

$$24 + 40 = \boxed{}$$

Ⓛ Test 9 Part 5 (Lesson 88, Ex. 9)

$$6 - 4 = \boxed{} \qquad 8 + \boxed{} = 13$$

$$7 + 3 = \boxed{} \qquad 26 + \boxed{} = 31$$

Remedies

Remedies CONTINUED

(M) Test 9 Part 6 (Lesson 83, Ex. 9)

$$40 + 9 = \boxed{}$$

$$30 + 7 = \boxed{} \qquad \boxed{} + \boxed{} = 25$$

(N) Test 9 Part 6 (Lesson 85, Ex. 9)

$$20 + 7 = \boxed{} \qquad \boxed{} + \boxed{} = 62$$

$$\boxed{} + \boxed{} = 58 \qquad 80 + 5 = \boxed{}$$

Connecting Math Concepts

Name _____

$8 + 5 =$ 13
|||||

$7 - 4 =$ 3
|||| ||||

$6 - 2 =$ 4
|||| ||

$7 + 4 =$ 11
||||

$$\begin{array}{r} 56 \\ +\ 14 \\ \hline \end{array}$$

$$\begin{array}{r} 39 \\ +\ 24 \\ \hline \end{array}$$

☆ TTTⱦ
|||||| ||| =

TTⱦⱦ
|||| ||| =

TTTT ⱦⱦⱦ |||||| =

Mastery Test 10

$9+1=$ ☐

$9+2=$ ☐

$12+1=$ ☐

$12+2=$ ☐

$17+1=$ ☐

$17+2=$ ☐

$8+1=$ ☐

$8+2=$ ☐

Independent Seatwork

☐ $-8=$

$39+$ ☐ $=$

| | | | | | | | | | | | |

Connecting Math Concepts

Remedies

Name _____

A Test 10 Part 1 (Lesson 99, Ex. 10)

$$7 - 5 = \boxed{}$$

$$15 + \boxed{} = 21 \qquad 8 + 3 = \boxed{}$$

B Test 10 Part 2 (Lesson 98, Ex. 7)

$$\begin{array}{r} 39 \\ + 24 \\ \hline \boxed{} \end{array} \qquad \begin{array}{r} 54 \\ + 32 \\ \hline \boxed{} \end{array}$$

C Test 10 Part 2 (Lesson 99, Ex. 8)

$$\begin{array}{r} 56 \\ + 14 \\ \hline \boxed{} \end{array}$$

Remedies CONTINUED

D Test 10 Part 3 (Lesson 99, Ex. 9)

ㅜㅜㅜㅜㅜ 〒〒〒||卌 =

ㅜ〒〒|||||||卌 =

E Test 10 Part 3 (Lesson 100, Ex. 9)

ㅜㅜ〒|||||卌 = | ㅜㅜㅜㅜ〒〒||||卌 =

F Test 10 Part 4 (Lesson 99, Ex. 7)

6+1= ☐

6+2= ☐

17+1= ☐

17+2= ☐

9+1= ☐

9+2= ☐

3+1= ☐

3+2= ☐

Remedies

Remedies CONTINUED

$$12 + 1 = \qquad 8 + 1 =$$

$$12 + 2 = \qquad 8 + 2 =$$

Remedies

Mastery Test 11

Name _____

$$\begin{array}{r} 6 \\ +40 \\ \hline \end{array}$$
$$\begin{array}{r} 1 \\ +27 \\ \hline \end{array}$$
$$\begin{array}{r} 3 \\ +70 \\ \hline \end{array}$$
$$\begin{array}{r} 1 \\ +35 \\ \hline \end{array}$$
$$\begin{array}{r} 7 \\ +50 \\ \hline \end{array}$$

56–12=
TTTTT
IIIIII

65–24=
TTTTTT
IIIIII

4–1=

9–1=

6–1=

3–1=

7–1=

39+10=

25+10=

74+10=

46+10=

54+10=

Mastery Test 11

$$45+25=\boxed{}$$

$$45-23=\boxed{}$$

$$\begin{array}{r} 27 \\ -\ 12 \\ \hline \end{array}\ \boxed{}$$

$$\begin{array}{r} 27 \\ +\ 34 \\ \hline \end{array}\ \boxed{}$$

$$\begin{array}{r} 9 \\ +\ 2 \\ \hline \end{array}$$

$$\begin{array}{r} 6 \\ +\ 2 \\ \hline \end{array}$$

$$\begin{array}{r} 13 \\ +\ 2 \\ \hline \end{array}$$

$$\begin{array}{r} 8 \\ +\ 2 \\ \hline \end{array}$$

$$\begin{array}{r} 4 \\ +\ 2 \\ \hline \end{array}$$

Independent Seatwork

$$\boxed{}-36=$$

TTT IIIIIIII

$$21+\boxed{}=$$

TTT IIII

$$59+\boxed{}=$$

T IIII

$$\boxed{}-16=$$

TTTTT
IIIIII

Remedies

(A) Test 11 Part 1 (Lesson 107, Ex. 8)

$$6 + 40 =$$

$$2 + 13 =$$

$$1 + 27 =$$

$$3 + 70 =$$

(B) Test 11 Part 1 (Lesson 108, Ex. 7)

$$\begin{array}{r} 2 \\ +\ 8 \\ \hline \square \end{array}$$

$$\begin{array}{r} 1 \\ +35 \\ \hline \square \end{array}$$

$$\begin{array}{r} 3 \\ +60 \\ \hline \square \end{array}$$

$$\begin{array}{r} 2 \\ +23 \\ \hline \square \end{array}$$

(C) Test 11 Part 2 (Lesson 101, Ex. 8)

$$65 - 24 =$$

TTT
TTT IIIII

(D) Test 11 Part 2 (Lesson 103, Ex. 8)

$$56 - 52 =$$

TTTTT
IIIIII

$$44 - 14 =$$

Remedies

Remedies CONTINUED

(E) Test 11 Part 3 (Lesson 109, Ex. 9)

3 – 1 =	6 – 1 =
4 – 1 =	7 – 1 =

(F) Test 11 Part 3 (Lesson 110, Ex. 9)

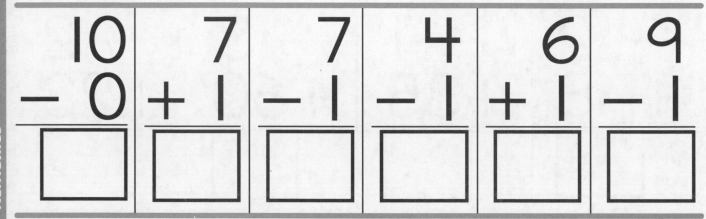

(G) Test 11 Part 4 (Lesson 106, Ex. 7)

(H) Test 11 Part 5 (Lesson 103, Ex. 9)

74+10 = ☐   46+10 = ☐

Remedies CONTINUED

I Test 11 Part 5 (Lesson 104, Ex. 7)

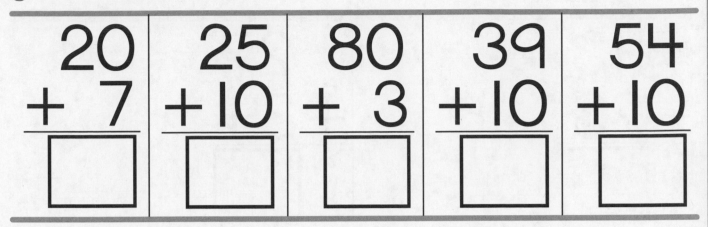

| 20
+ 7
☐ | 25
+10
☐ | 80
+ 3
☐ | 39
+10
☐ | 54
+10
☐ |

J Test 11 Part 6 (Lesson 108, Ex. 6)

27
−12
☐

46
+14
☐

45−23= ☐

Remedies

Remedies CONTINUED

(K) Test 11 Part 6 (Lesson 110, Ex. 6)

$$43 - 23 = \boxed{}$$

$$27 + 34 = \boxed{}$$

$$45 + 25 = \boxed{}$$

(L) Test 11 Part 7 (Lesson 102, Ex. 7)

$$9 + 2 = \boxed{}$$

$$13 + 2 = \boxed{}$$

$$8 + 2 = \boxed{}$$

$$16 + 2 = \boxed{}$$

Mastery Test 12

Name _____

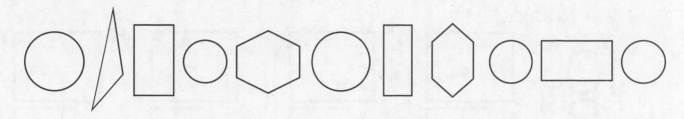

	1	2	3	4	5	6
shapes						

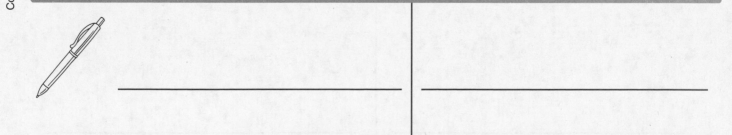

Mastery Test 12

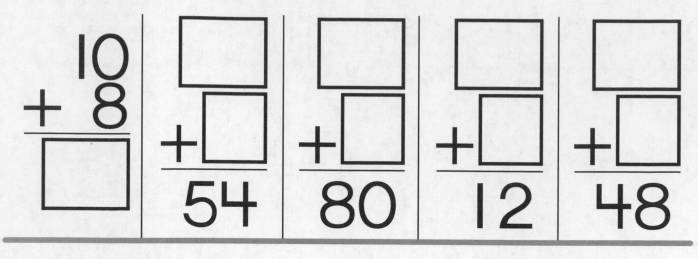

$$\begin{array}{r} 10 \\ + \ 8 \\ \hline \square \end{array}$$

| 1 | 2 | 5 | 4 | 8 | 7 | 6 | 3 |

3−1 =   7+1 =   7−1 =

8+2 =   5−1 =   5+2 =

Independent Seatwork

56+24 = ☐   44−31 = ☐

Remedies

A Test 12 Part 1 (Lesson 120, Ex. 11)

shapes	1	2	3	4	5	6
△ ▯ ○						

B Test 12 Part 2 (Lesson 119, Ex. 7)

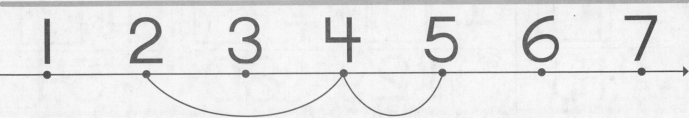

$$\frac{2}{4-}$$

$$\frac{4+}{5}$$

C Test 12 Part 3 (Lesson 116, Ex. 7)

Remedies

Remedies CONTINUED

D Test 12 Part 3 (Lesson 117, Ex. 9)

_____ | _____

E Test 12 Part 4 (Lesson 115, Ex. 8)

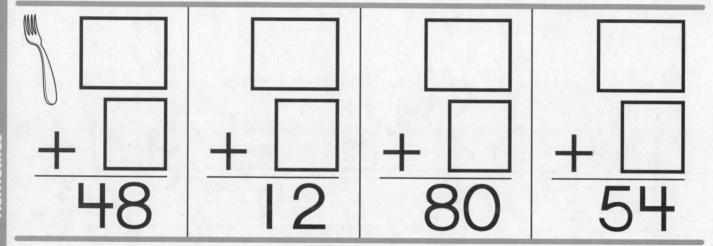

F Test 12 Part 5 (Lesson 116, Ex. 8)

3   4 __ __ 6 __ __ 8 __ __ 10 __

G Test 12 Part 5 (Lesson 118, Ex. 7)

14 17 16 15 18 19 20

Remedies CONTINUED

(H) Test 12 Part 6 (Lesson 113, Ex. 6)

8+2=	4−1=	9+1=
7+2=	10−1=	5+2=

(I) Test 12 Part 6 (Lesson 114, Ex. 7)

2+10=	7−1=	63+2=
7+1=	2−1=	17+2=
63+10=	3−1=	17+10=

(J) Test 12 Part 6 (Lesson 115, Ex. 7)

18+10=	8−1=	35+2=
5−1=	8+2=	10−1=
58+10=	18+1=	5+10=

A-1

☐   TTT   ☐   TTTTT   ☐   IIIII   ☐   TTTT

A-2

☐   TTIIIIIII   ☐   TIII   ☐   TTTTTIII   ☐   TTTT TTTT IIIII

A-3

☐ ☐ ☐ ☐ ☐ ☐

A-4

20	40	7	30

A-5

35	24	15	41

8+1 = □   5+1 = □ $\begin{array}{r} 12 \\ +\ 1 \\ \hline \square \end{array}$ $\begin{array}{r} 12 \\ +\ 2 \\ \hline \square \end{array}$

8+2 = □   5+2 = □

B-6

7+2 = □ $\begin{array}{r} 6 \\ +\ 2 \\ \hline \square \end{array}$ $\begin{array}{r} 9 \\ +\ 2 \\ \hline \square \end{array}$ $\begin{array}{r} 14 \\ +\ 2 \\ \hline \square \end{array}$

3+2 = □

B-7

B-8

B-9   7−1 =   3−1 =   9−1 =

6+1 =   6−1 =   2+2 =

10−1 =   2+2 =   9+2 =

B-10

30+10 =   50+10 = $\begin{array}{r} 27 \\ +10 \\ \hline \square \end{array}$ $\begin{array}{r} 49 \\ +10 \\ \hline \square \end{array}$

34+10 =   58+10 =

B-11   B-12

Connecting Math Concepts

$$70 + 8 = \boxed{}$$

$$\boxed{} + \boxed{} = 52$$

B-13

$$\begin{array}{r} \boxed{} \\ + \boxed{} \\ \hline 46 \end{array}$$

$$\begin{array}{r} 30 \\ + \ 9 \\ \hline \boxed{} \end{array}$$

$$10 + 4 = \boxed{}$$

$$\boxed{} + \boxed{} = 60$$

B-14

$$\begin{array}{r} 12 \\ + \ 2 \\ \hline \boxed{} \end{array}$$

$$\begin{array}{r} \boxed{} \\ + \boxed{} \\ \hline 13 \end{array}$$

 =    =   (quarter) =   (nickel) =

C-15

(6 pennies) =   (3 dimes) =

(5 pennies) =   (nickel + 4 pennies) =

C-16

 =    =

C-17

| 21 | 19 | 58 | 60 | 107 | 111 | 82 | 78 |

C-18

50 FIFTY DOLLARS	5 FIVE DOLLARS	1 ONE DOLLAR	1 ONE DOLLAR	1 ONE DOLLAR	=
20 TWENTY DOLLARS	10 TEN DOLLARS	10 TEN DOLLARS	10 TEN DOLLARS	5 FIVE DOLLARS	=
10 TEN DOLLARS	10 TEN DOLLARS	5 FIVE DOLLARS	1 ONE DOLLAR	1 ONE DOLLAR	=

C-19

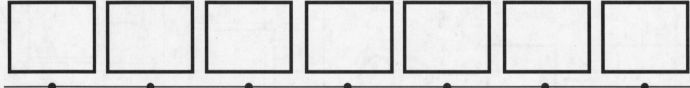

5   8   7   6   11   10   9

C-20

$6 + \boxed{} = 10$ $12 + \boxed{} = 15$

D-21

$8 + 3 = \boxed{}$ $2 + \boxed{} = 7$

$7 - 2 = \boxed{}$ $5 + \boxed{} = 8$

D-22

Connecting Math Concepts

$36 + 40 =$
TTTT

$36 + 4 =$
||||

$25 + 3 =$
|||

$25 + 30 =$
TTT

D-23

$42 + 23 =$
TT |||

$37 + 43 =$
TTTT |||

D-24

$61 + 21 =$

$58 + 34 =$

D-25

$57 - 23 =$
TTT T̶ ||||| H̶

$44 - 12 =$
TTT T̶ || H̶

E-26

$65 - 34 =$
TTTTTT |||||

$54 - 21 =$
TTTTTT ||||

E-27

$36 - 22 =$

$43 - 13 =$

E-28

$$24$$
$$+15$$
$$\overline{}$$

$$46$$
$$-32$$
$$\overline{}$$

$$34-31=$$

$$34+31=$$

E-29

_____

E-30

$$87-69=18$$

$$47+15=62$$

_____

$$64+25=89$$

$$52-21=31$$

Connecting Math Concepts

$$\frac{6}{8} \quad = \frac{7}{\ } \quad \frac{7}{10}$$

F-32

KMLMKMLPMLM

K =   L =   M =   P =

Letter	1	2	3	4	5	6

G-33

 A   B   C   D   E   F

G-34